**THE FIRE PREVENTION AND
INFORMATION BOOKLET FOR HOMES**

SAFEBOOK

...

YOUR FAMILIES RESOURCE FOR FIRE CAUSES, FIRE SAFETY, FIRE HAZARDS AND FIRE PREVENTION

BY DERRICK WHITEHEAD, C.E.O. AND C.F.F.
Credit: Pleasureofart

AuthorHouse™
1663 Liberty Drive
Bloomington, IN 47403
www.authorhouse.com
Phone: 833-262-8899

Because of the dynamic nature of the Internet, any web addresses or links contained in this book may have changed
since publication and may no longer be valid. The views expressed in this work are solely those of the author and do
not necessarily reflect the views of the publisher, and the publisher hereby disclaims any responsibility for them.

Any people depicted in stock imagery provided by Getty Images are models,
and such images are being used for illustrative purposes only.
Certain stock imagery © Getty Images.

This book is printed on acid-free paper.

ISBN: 978-1-6655-7829-5 (sc)
ISBN: 978-1-6655-7828-8 (e)

Library of Congress Control Number: 2022923305

Print information available on the last page.

Published by AuthorHouse 02/23/2023

HOUSE

NOTES FROM THE EDITOR

By Derrick Whitehead

SafeBook is designed with your family's safety in mind. As a certified firefighter and founder of the Office Fire Prevention Inc. it is with my greatest of intentions to share with all my friends' images and information about home fire causes, home fire safety, home fire hazards, home fire injuries and home fire prevention. SafeBook is about saving lives and this I hope that you can admire as you learn more about the danger of and the ways to detect and combat (when reasonable) fires. Fire is a widespread tragedy that each year kills approximately 4,000 people and injures about three times as many. Seniors and Children are particularly vulnerable to fire; therefore, it is important that families have the tools (of Detection and Protection) and know the rules of fire safety.

My hopes are that all families will receive the information enclosed with PRIDE (Parent Resource for Information, Detection and Education) in their efforts to exercise safety by allowing SafeBook to be their guide. This booklet will be instrumental for families and children. In order for our communities to be a safer place to live, work and play, families are encouraged to exercise safety first at all times and in all they do because safety first is not just a slogan it's an actionable attitude. As a reward for you and your family's commitment to Safety First, the Office of Fire Prevention Inc. wants to make a Safety Advocate out of you and as you read further in this book, you learn exactly what you will need to do.

DEDICATION

SafeBook is dedicated to those unselfish members of the community that hold devotion to responsibility above personal beliefs, those members who exercise sincerity of the safety situation above convenience, those members of the community who strive for better ways to protect themselves, their family and the community from the ravage of fire and other disasters that threaten life. This book is dedicated to…

THE SAFETY ADVOCATE IN YOU

THE SAFEBOOK SAFETY PLEDGE

TODAY IS A GREAT DAY FOR A SAFE DAY IN MY HOME AND COMMUNITY. TODAY I WILL PRACTICE AND PROMOTE SAFETY BEHAVIOR WHERE EVER

FIRE/COMBUSTION

Fires are the fifth leading cause of accidental deaths in the US, and unfortunately most people have no knowledge of this information or choose to ignore it. Local fire departments annually respond to approximately 1, 300,000 fire calls, resulting in roughly 4,000 deaths each year, 15,000 civilian injuries and $20.0 billion in direct property damage.

It is imperative that families learn the danger and destruction that fire causes. A fire can be a devastating experience and learning the necessary methods of prevention, broadens your window for preservation.

Fire is a by-product of a larger process called combustion. Fire and combustion are two words used interchangeably by most people, especially firefighters.

Fire can be a rapid process that quickly destroys an entire structure. This mostly occurs when there are many combustibles in the area (i.e., paper, wood, rubber, and plastic). Fire burns in two modes flaming or surface combustion. Flaming combustion is the burning of logs in a fire fireplace. Surface combustion is the continued burning of an object, which has no visible flames.

HOW FIRES START

Fire is a chemical reaction involving rapid oxidation or burning of a fuel.
It needs four elements to occur.

- **Fuel**
  Fuel can be any
  combustible material-solid
  liquid or gas. Most solids and
  liquid becomes a vapor or gas
  before they will burn

- **Oxygen**
  The air we breathe
  is about 21 percent oxygen. Fire
  only needs an atmosphere with at
  least 16 percent oxygen

- **Heat**
  Heat is the energy
  necessary to increase the
  temperature of the fuel to a point
  where sufficient vapors are given
  off for ignition to occur

- **Chemical Reaction**
  A chain reaction can occur when other
  three elements are present in the proper
  conditions and proportions. Fire occurs
  when this rapid oxidation or burning takes place.

Take any of these factors away, and the fire cannot occur or will be extinguished if it was already burning.

CHILDREN AND THE COMMUNITY

The community that we live in has a significant impact on our lives. Families should make it their business to learn about their community. Children are influenced by sentiments, opinions, and moral standards that exist not only in the home, but also in their environment. For children Imitation comes first and comprehension later.

Communication with children is a vital part of learning about their day-to-day activities and hobbies. Unfortunately, our children are exposed to all sorts of danger and abuse which has the ability to influence their safety behavior. One of the effects to our children is the development of fire starters. Children become fascinated with fire as early as their toddler stage. Children become acquainted with fire when they are around smokers and other children with matches and or cigarette lighters. The flame of a match or cigarette lighter is colorful, exciting, and mysterious.

Most children do not understand, in a split second of ignition, a fast-moving fire can scar them for life or kill them and their families. Most of the children who die in fires each year in the United States die in fires set by themselves or other children who are fire starters in the community. In 2021, children playing with fire caused an estimated 284 deaths and 2,158 injuries.

Playing with matches or lighters causes about three out of four fires set by children. Children start 100,000 fires each year.

FOUR MAJOR COMPONENTS
OF AN EVACUATION PLAN

Two ways out of each room

A meeting place outside

Working smoke alarms

Practice Practice Practice

RULES OF FIRE SAFETY

1. Smoke detectors save lives. They give people time to get out before a fire gets too hot and smokey. Smoke and heat are the major killers, not flames.

2. Have an escape plan. Practice it monthly so it becomes second nature. Children get frightened and need to know there is another way out if the path to the front door is blocked. Try to use another door, drop out of a first-floor window, or have a collapsible ladder that can be put over the sill of a bedroom window. An important safety tip: check all windows, especially after painting to make sure they open easily.

3. Sleep with the bedroom door closed. It keeps out smoke and flames for as much as 20 minutes

4. Roll out of bed and crawl under the smoke. Do not stand up because that will put your head where the heaviest smoke is.

5. Do not hide in closets, under the bed or in a bathtub.

6. Know your way out. Flames might block the main route.

7. Feel the door. Use the back of your hand; its mor sensitive. If the door is not hot, open it. If it is hot, do not open it. If there's smoke, close it again and try a window. Do not jump from upper windows. Put a sheet or pillowcase or something at the window so firefighters can be alerted that someone is in the room.

8. Do not go back. Go to your family meeting location. A meeting place across the street is best, in case the fire spreads to other buildings... and **look before crossing!**

9. Call 911 for help. Call after you leave the building. Do not assume someone else has called.

10. Practice your escape plan. It is very important that small children remember the plan. The rules are different if you live in a high-rise building. Unless the fire is on the first floor, put wet towels along the door edges and go to a window to signal your presence, but stay in your own unit.

ONE FAMILY AND TWO FAMILIES HOUSE ESCAPE PLAN

1. When you first hear the alarm, roll off your bed and crawl onto the floor. Do not sit up in your bed. If there's smoke in the room, sitting up in bed or standing up may put you at a level where it is dangerous to breathe or expose you to tremendous heat that will instantly debilitate you. By staying low, you give yourself an opportunity to survive in a room that could otherwise be your death.

2. Crawl to your bedroom door and feel it with the back of your hand. Do not collect any valuables or clothes. You and your family are the only valuables worth saving. If the door is hot to the touch, don't open it. It could mean that the fire is just outside of your door. Opening the door could bring a rush of flames into your room. Alternatively, opening the door could bring a gust of fresh oxygen to the flames and create a backdraft – the explosive fire that ignites everything and everybody in its path. A closed door makes a strong defense against fire and can give you protection until help arrives.

3. If the door is hot, seal the cracks around the door. Wet towels and blankets works best, but in an emergency, newspapers, clothes, or sheets will do.

4. Next, crawl to a window and open it slightly at the top and bottom. Opening a window all the way might pull smoke into your room from outside the door.

5. Break open a window only as a last resort. If you break a window and smoke is coming in from outside the house, there will be no way to close it. Also, a broken window may bring of rush of smoke from inside the home.

6. If you are on the first floor and can safely escape through the window, crawl out to safety. Above the first floor, don't consider jumping except as an absolute last resort. Many people who could have survived in a room while waiting for help to arrive have died by jumping from dangerous heights.

7. If you can't get out of your room, once you're at the window, make noise to let others know you are trapped inside. Banging objects together is better than shouting because it's important to save your breath. Waving a towel or sheet out the window will help signal your location.

8. If there is a phone in your room, call the fire department or 911 to report the fire and let them know that you are trapped inside.

9. Stay low by the window. This will allow you to breathe fresh air from the outside while avoiding the smoke that is likely to be seeping into your from inside the house. Don't forget that poisonous carbon monoxide is colorless and odorless, so don't expect to see it or smell it.

10. If your bedroom door is hot, open it sightly to look for smoke or flames.

11. After you leave your room, close the door. This will prevent the spread of the fire and help protect your property.

12. If the path looks clear, crawl along your predetermined escape route. Pound along the walls as you crawl and shout "Fire!" to alert others.

13. If possible, place a wet cloth over your nose and mouth before escaping. No matter what, stay low.

14. Don't look for the fire. The precious time you spend searching for a fire may be the same moments you need to escape the home.

15. Unless you are trapped in a room with a phone, don't take time to call the fire department before escaping. Many people have died while calling the fire department. Remember, you don't have time in a fire.

16. Go directly to your family's predetermined meeting place outside of the home. Don't wait for everyone to meet in the house before escaping. Once you are outside, call the fire department (never assume that it has been done by a neighbor). Stay on the line until the department has all the information that need. If you are at an alarm box, stay by the box in case you need to direct the fire truck to the fire.

17. Don't go back into the house until the fire department determines it's safe to return.

HOW FIRES KILLS

(Something you need to know)

- **How do most people die in fires?**

One quarter of a home fire victim's die from burns, seven out of ten dies from breathing poisonous gases produced by the fire.

- **What are deadly gases?**

The most common deadly gas produced by fire is carbon monoxide (produced by all burning items). Other gases include hydrogen cyanide (from burning wool silk nylon and some plastics). Hydrogen chloride (detected by its pungent odor and irritation to eyes) and carbon dioxide (a gas given off by the fire itself gets into the lungs and effects the respiratory system). Enter CARBON MONOXIDE DECTECTOR

- **What happens to the oxygen?**

Fire consumes oxygen, reducing the airs normal (21 percent) oxygen level. When the atmospheric oxygen drops below 17 percent, people get extremely disoriented. When it drops between 6 and 10 percent, people cannot breathe. Hot air and smoke rise, and some deadly gases sink to the floor. Remember when in a smoked filled room, crawl with your head in safety zone 1 to 2 feet above the floor. ENTER SMOKE DETECTOR

"A BURNING ISSUE"

DID YOU KNOW???

House fires are deadly. In fact, 4,500 people die in them in the U.S each year. Sadly, homeowners contribute to that toll with their misconceptions about escape time. These errors in judgement are especially deadly when compounded by the general lack of preparedness for fire.

According to a recent poll by the Quincy, Massachusetts-based National Fire Protection Association,1/4 of the five hundred people surveyed wrongly, believe it would take ten minutes or more for a fire to create life threatening conditions, when they can occur in 1/5 that time. More realistically, 42% of people believe that a living room fore could produce life- threatening conditions in a second-floor room in less than two (2) minutes.

How quickly can a fire spread? The time -lapse graph shows how a fire develops from ignition in a second-floor couch. The graph shows that smoke from that fire is not evident to someone outside the house until almost five minutes have elapsed. What more a smoke detector might not sound an alarm for almost two (2) minutes, dangerously the analysis and research stress the need for higher public awareness. "Until people become more knowledgeable about the real potential for fire, we will continue to suffer each year," Hall says.

Understanding just how quickly a fire can spread and creating an escape plan are great ways to start- J.D.W

30-Fire ignites and grows rapidly
1:04-From the first flame fire spreads and smoke begins to fill the room
1:35-Smoke layer descends rapidly, temperature exceeds 190 degrees
1:50- Smoke detector at foot of stairs sounds, still time to get out
2:30-Temperature above couch over 400 degrees
2:48-Smoke pours into other rooms
3:03-Temperature three ft above floor in a room of origin over 500 degrees (NO ONE COULD SURVIVE)
3:20-Upstairs Hall filled with smoke, escape more difficult
3:41-FLASHOVER, Energy in room of origin ignites everything
(TEMPERATURE IS 1,400 DEGREES)
3:50-Two minutes after smoke detector sounds, second exit only way out
4:33- Flames just now visible from exterior of house. First evidence of fire from outside
AT THIS POINT RESCUE MIGHT NOT BE POSSIBLE
42% of people believe a living room fire could produce life-threatening conditions in a bedroom in two minutes or less- that is realistic.
58%believe it would take longer than two minutes for conditions to become deadly in an upstairs bedroom
24%believe they will have ten minutes or more before life threatening conditions develop: this is very unrealistic and suggests a potential deadly lack of urgency.

PHYSIOLOGICAL EFFECTS
OF REDUCE OXYGEN

Oxygen in Air (Percent)	Symptoms
21	None – Normal conditions
17	Some impairment of muscular coordination: Increase in respiratory rate to compensate for lower oxygen content.
12	Dizziness, headache, rapid fatigued
9	Unconsciousness
6	Death within a few minutes from respiratory failure and concurrent heart failure.

Note: These data cannot be considered absolute because they do not account for Difference in breathing rate or length of time exposed.

These symptoms occur only from reduced oxygen. If the atmosphere is contaminated with toxic gases, other symptoms may develop.

PHASES OF A FIRE AS IT RELATES TO THE LEVEL OF OXYGEN

INCIPIENT PHASE

- Room temperature slightly over 100°F (38°C)
- Rising hot gases
- Room air approximately 20% oxygen

ROLLOVER POTENTIAL

- Free-burning fire
- Smoke and superheated gases collecting at ceiling level

ROLLOVER

- Superheated vapors ignite
- Flame front rolls across ceiling

STEADY-STATE PHASE

- Room temp. approx. 1,300°F (700°C)
- Heat accumulating in upper areas
- High oxygen supply
- Full fire involvement

FLASHOVER

- Simultaneous ignition of all combustibles in room
- High heat level from floor to ceiling

BACKDRAFT POTENTIAL

- Puffs of smoke leaving building
- Pressurized smoke exiting small openings
- Black smoke becoming dense gray-yellow
- Rapid inward suction of air when opening is made
- Muffled sounds
- Little or no visible flame
- Smoke-stained windows

HOT-SMOLDERING PHASE

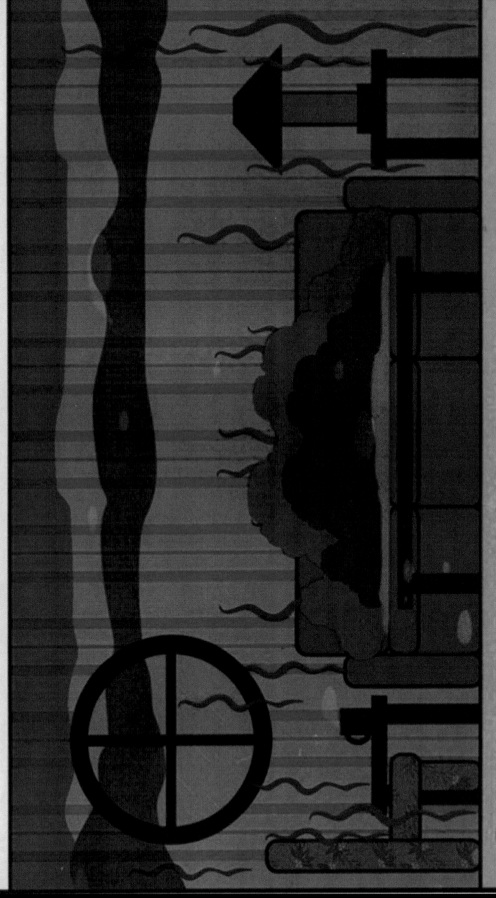

- Room temperature high throughout
- High carbon monoxide and carbon levels in heavy smoke
- Oxygen below 15%

"WORKING TOGETHER TO SAVE LIVES THROUGH DETECTION WHICH INCREASES THE CHANCE OF PROTECTION".

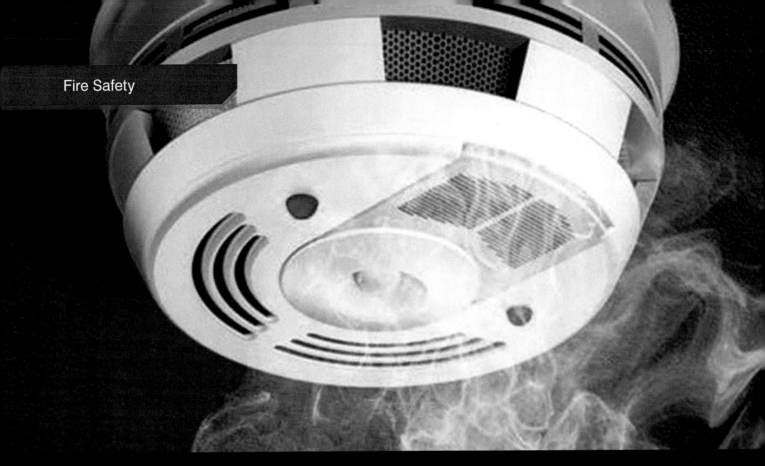

SMOKE DETECTOR

A smoke detector is a sensor that detects smoke as a primary indication of fire. It provides a signal to a fire alarm system in a large building or produces an audible and visual signal locally in a room or a home. Smoke detectors are usually housed in a small, round shaped plastic case, and placed at the roof where there are risks of fire or fire hazards.

They should be placed inside each bedroom, outside each sleeping area and on every level of the home, including basement. On levels without bedrooms, install alarms in the living room or near the stairwell to the upper level or in both locations.

There are two main types of smoke detectors: photoelectric and ionization. When smoke enters the detector, a photoelectric type detects sudden scattering of light, whereas an ionization type detects the change of electrical current flow that triggers the signal – indicating the presence of smoke.

Smoke detectors have an average life of about 10 years. Detectors need to be tested periodically and the batteries changed when required (at least once a year or when it starts to chirp).

SMOKE DETECTORS

A fire-protection device that automatically detects and gives a warning of the presence of smoke, increasing the famalies chances of safety and survival in a fire situation.

TYPES OF SMOKE DETECTORS:

Some detectors operate on one of two basic principles; ionization or photoeletric. For maximum protection, the user should understand the advantages and disadvantages of both types.

IONIZATION

The ionization detector used a small amount of radioactive material to make the air within a sensing chamber conduct electricity. When very small smoke particles enter the sensing chamber, they interfere with the conduction of eletricity, reducing the current and triggering the alarm. The particles to which the detectors respond are often smaller that can be seen with the human eye. Because the greatest number of these invisible particles are produced by flaming fires, ionization detectors respond slightly faster to open flaming fires than do photoelectric detectors. The radiation sources in ionization detectors is not a hazard to the home's occupants.

PHOTOELECTRIC:

The photoelectric detector uses a small light source- either an incandescent bulb or a light-emitting diode (LED)- that shines its light into the dark sensing chamber. The sensing also contains an electrical, light- sensitive component known as a photocell. The light source and photocell are arranged so that light from the source does not normally strike the photocell. When smoke particles enter the sensing chamber of the photoelectric detector, the light is reflected off the surface of the smoke, allowing it to strike the photocell and increase the voltage from the photocell.

POWER SOURCES:

Batteries or household current can power residential smoke detectors. Battery- operated detectors offer the advantage of easy installation with the use of a screwdriver and a few minutes is all that is needed. Battery models are also independent of house power circuits and will operate during power failures. It is crucial that only the specific battery recommended by the detector manufacturer be used for replacement. In many circumstances, a unit with the wrong battery will not respond to smoke even though its test button will function. Additionally, some detectors require special batteries that may be difficult to find in all but large metropolitan areas or are available only by mail order. The batteries should be changed at least twice a year or more if necessary. An effective way to remember when to change the batteries is in the spring and fall when the clocks are set forward and back or when it chirps.

SMOKE DETECTOR PLACEMENT

CORRECT AND INCORRECT POSITIONING FOR SMOKE DETECTORS

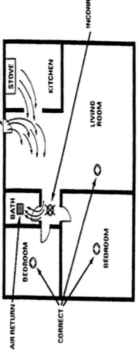

Source: First Alert

POSITIONING THE SMOKE DETECTOR

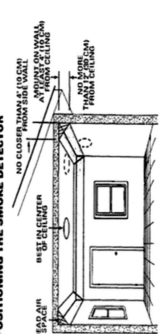

Source: First Alert

The smoke detector should be assessed weekly according to the manufacturer's instructions. It is important that owners assess their detectors at least once a month with smoke aerosol product specifically designed for this purpose, because some detectors have test buttons that only check the devices horn circuit. The owner the detector may not realize this. Only when the detector incorporates the test button that stimulates smoke or checks the detector's sensitivity can the "smoke test" be eliminated. Smoke detectors with test buttons that stimulate smoke or check sensitivity are recommended in homes were "smoke testing" is not likely to be conducted such as a handicapped or elderly person's home.

FIRE EXTINGUISHER

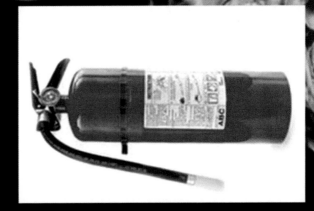

FIRE EXTINGUISHERS CAN SAVE LIVES AND PROPERTY BY PUTTING OUT SMALL FIRES OR CONTAINING THEM UNTIL THE ARRIVAL OF THE FIRE DEPARTMENT!

Fire Extinguishers are designed for

➤ Small fires

➤ Confined fire areas

➤ People who have read the directions in advance and know how to operate

Fire Extinguishers are NOT designed for

➤ Large fires

➤ Growing fires

➤ People who are unfamiliar with directions and do not know how to operate

30

KNOW THE PASS WORD OF
FIRE EXTINGUISHERS

P—PULL THE PIN — Some Fire Extinguishers require releasing a latch, or pin. This allows discharge of the unit.

A—Aim at Base of Fire — Point horn, hose, or nozzle at the base of the fire. Hold unit upright. Stand 6-8 feet away.

S—SQUEEZE HANDLE — This discharges the unit. Some units have a button to press. Releasing the handle or button will STOP the discharge.

S—SWEEP SIDE BY SIDE- Move in towards the fire. After fire is out, WATCH AREA. If it reignites, repeat the process.

- ✓ Some fire extinguishers discharge in as little as eight seconds.
- ✓ Check your own unit for specific instructions.
- ✓ Check it monthly

KNOW THE A B C, S OF FIRE CLASSES

A – Ordinary Combustibles
- ✓ Wood
- ✓ Paper
- ✓ Plastic
- ✓ Cloth

B – Flammable Liquids
- ✓ Grease
- ✓ Oil
- ✓ Flammable Liquids

C – Electrical Equipment
- ✓ "C" Indicates it has a non-conducting agent. It is used for electrical fires.
- ✓ NEVER USE WATER ON THESE FIRES!!!!!!!

D – Flammable Metals

Warning: Using a fire extinguisher on the wrong class of fire **CAN MAKE THE FIRE WORSE.**

NUMBER RATING: Indicates the most area a fire extinguisher can put out.

A RED SLASH: A slash across any symbol indicates the fire extinguisher is NOT intended for that class of fire. Missing symbols indicates the unit has not been assessed for that type of fire.

BEFORE A FIRE OCCURS KNOW YOUR FIRE EXTINGUISHER

Location
- ✓ Fire extinguishers should be in plain view
- ✓ Within an easy reach, but away from children
- ✓ Near escape routes
- ✓ Do you know where your fire extinguisher is located?

Operation
- ✓ There should be a handle or lever for carrying and discharging
- ✓ The safety pin locks to prevent discharging
- ✓ Most have a gauge to tell if the unit is full
- ✓ A discharge nozzle may be long, short or a hose
- ✓ There should be instructions on the label
- ✓ The class of fire extinguishers should be on the label

Inspection & Care
- ✓ Make sure the safety lock has not been broken
- ✓ Check gauges: GREEN is not good reading; RED means replace or recharge
- ✓ Disposable unit: Are no longer good after any discharge or broken safety pin; one time use; check them monthly
- ✓ Rechargeable units: Recharge after every use and any use. Have unit checked professionally every year. Have unit pressure assessed every five years.

FIRE EXTINGUISHERS CAN SAVE LIVES AND PROPERTY BY PUTTING OUT SMALL FIRES OR CONTAINING THEM UNTIL THE ARRIVAL OF THE FIRE DEPARTMENT!

Fire Extinguishers are designed for
- ➢ Small fires
- ➢ Confined fire areas
- ➢ People who have read the directions earlier

Fire Extinguishers are Not designed for
- ➢ Large fires
- ➢ Growing fires
- ➢ People who are unfamiliar with directions

BEFORE A FIRE OCCURS KNOW YOUR EXIT

EXITS should be:
- ➢ Clear and open pathways
- ➢ Known to all those in the household
- ➢ Doors that open easily
- ➢ Clear and concise-know who may need additional assistance in the household

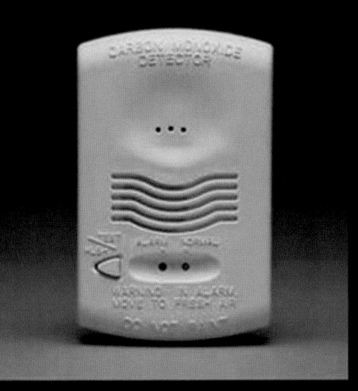 

CARBON MONOXIDE DETECTOR

Carbon monoxide detectors work much like your fire or smoke alarm by sounding a siren when they detect carbon monoxide.

Carbon monoxide detector should be installed on each level of the home, especially on the levels with fuel burning appliances (furnace, water heater and fireplace) and outside of sleeping areas.

Carbon monoxide is a colorless, odorless, and poisonous gas that claims over 430 lives a year.

Carbon monoxide poisoning typically occurs from breathing in carbon monoxide at excessive levels. Symptoms or often described as "flu-like" and commonly include headache, dizziness, weakness, vomiting, and shortness of breath.

It's a byproduct of burning carbon fuel like the natural gas in your stove and the gasoline in your car. Even small doses of carbon monoxide can cause permanent damage or death.

CARBON MONOXIDE DETECTOR PLACEMENT

Only half of U.S. homes have a working carbon monoxide alarm.

75% of homes have a potential source of carbon monoxide.

72,000 CO incidents occur each year in the U.S.

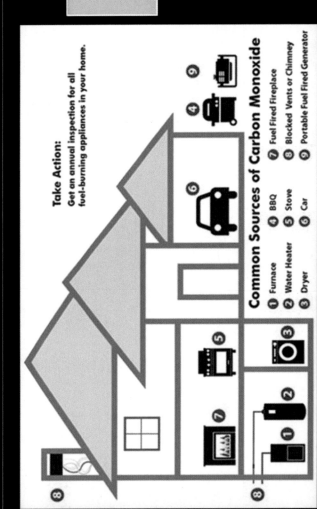

Take Action:
Get an annual inspection for all fuel-burning appliances in your home.

Common Sources of Carbon Monoxide

1 Furnace
2 Water Heater
3 Dryer
4 BBQ
5 Stove
6 Car
7 Fuel Fired Fireplace
8 Blocked Vents or Chimney
9 Portable Fuel Fired Generator

CARBON MONOXIDE

THE LAW STATES THAT YOU MUST HAVE A Carbon Monoxide Detector!

What is Carbon Monoxide (CO)?

It is an order less gas produced by burning fuel. Exposure to lower levels of **CO** over several hours can be just as dangerous as higher levels for a few minutes of exposure.

Proper Ventilation:

Venting Gas Range
Never Use Gas Range for Heating
Never Leave Cars Running in an Open or Closed Garage
Start Yard Equipment Outside
Never Use Charcoal of Gas Grill Indoors
Vent Fuel Burning Heaters

People At Risk

Children
Elderly
People with Lung Disease
People with Heart Disease
Pregnant Women

Signs & Symptoms

Headaches: Fatigue
Sleepiness
Weakness
Nausea: Vomiting
Dizziness: Confusion
Tightness in the Chest
Trouble Breathing
LOSS OF CONSCIOUSNESS
COMA
DEATH

Dangerous Levels Can Occur If:

Improperly Installed
Improperly Maintained
Damaged
Malfunctioning
Improperly Used

What to Do:

Ventilate the house! Get Out!
Turn off fuel burning appliances, if possible, as you leave
Get fresh air!
Call 911
Seek medical attention even if you feel better!
Have your home checked by a professional!
Do not go back in until the problems have been corrected!

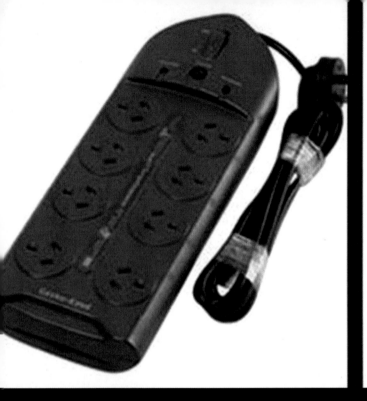

SURGE PROTECTOR

A surge protector is an electrical device that protects electronic devices from voltage spikes (voltage fluctuations) and other electrical irregularities. Surge protectors redirect the extra energy into the ground wire, preventing any damage to connected devices.

Most people don't know that electronic devices can be damaged by voltage spikes and other electrical irregularities. Even if your appliances are turned off, a surge protector keeps them safe from sudden fluctuations in voltage and other power anomalies.

If you use surge protectors, you also need to be absolutely sure that they have internal overload protection. And just because the strip has six plugs in it does not mean that you necessarily want to plug six things into it.

And whether you are using surge protectors or extension cords you should avoid plugging heavy-duty appliances into them. This includes things like microwaves and space heaters.

The term "common" could be misleading to some individuals. It refers to the probable frequency of the hazard being found and not to the severity of the hazards. **A common fire hazard is a condition that is prevalent in all occupancies and will encourage a fire to start.** Common fire hazards are listed in contrast to special hazards, which are usually characteristic of a given industry. Some common hazards are poor housekeeping, heating, lighting, and power equipment, floor cleaning, compounds, packing materials, fumigation substances, and other flammable and combustible liquids.

Families should be alert for the dangers posed by these common hazards. Poor housekeeping can make maneuvering difficult through a home or work station. **Poor housekeeping also increases the fire load in an area and increases the chance that flammable or combustible material may encounter an ignition source.** It also hides fire hazards in the clutter. Improperly functioning heating, lighting, or other electrical equipment can provide an ignition source for nearby combustibles liquids that are improperly used and stored can provide a volatile source should an ignition source be present.

Families must also make sure that exits are adequate in number and in reasonable condition. This can be accomplished by checking to make sure that all storage is removed from exit doors (especially those equipped with panic hardware) and make sure that any chains, deadbolts, or other extra locking devices are moved immediately.

Personal hazards are the most serious of all hazards. The term personal hazard covers all individual traits, habits and personalities of the people who work, live, or visits your dwelling, property or business. Personal hazards may be considered intangible, but they are always present. Good fire prevention practices can minimize personal hazards.

A fire hazard may be different as a condition that will encourage a fire to start or will increase the extent or severity of the fire. To prevent a situation from becoming hazardous, the fuel supply heat source, and the oxygen supply must be considered. If any one of these can be eliminated, a fire cannot occur. The oxygen supply hazard is normally present in air, and very little control can be maintained over the oxygen supply except in special cases.

Heat Source Hazards:

- **Chemical heat energy-** Heat of combustion, spontaneous heating, heat of decomposition, heat of solution. This occurs because of material being improperly used or stored. The materials may meet each other and react, or they may decompose and generate heat
- **Electrical heat energy-** Resistance heating, di-electric heating, heat from arcing, heat from static electricity. Poorly maintained electrical appliances, exposed wiring, and lighting are sources of electrical heat energies.
- **Mechanical heat energy-** Friction heat and heat of compression. Moving parts on machines such as belts and bearings, are a source of mechanical heating

Items that determine the Safety of the Home, Apartment, or Condominium

WINDOWS

- Repair missing and boarded up windows
- Replace broken and cracked windows. Seal drafty and loose trim.
- Install window locks on all basement and first floor apartment windows.
- Windows must operate properly. missing chains, handles, sashes, or windows that are painted shut will result in fail rating.

ELECTRIC

- Fixtures must work if present.
- Replace missing or broken cover plates, pull chains and exposed frayed and cracked wire.
- Two electric outlets or one permanent light fixture and one electric outlet are required in each room.

SMOKE DETECTORS

- There must be at least one working smoke detector for each living room or enclosed heating unit.

CARBON MONOXIDE DETECTORS

- Not less than one carbon monoxide detector is required in every dwelling if gas, coal, fuel oil, propane, kerosene, etc. is burned for heat. Carbon monoxide detectors must be installed with 40 feet of all rooms.

DOORS

- Replace missing doors,
- handles and locks.
- Repair inoperable locks, loose or missing thresholds, broken doors, and windows and doors that are out of adjustment. Closet doors also included in this category.

HANDRAILS

- Handrails are required for flights of four or more stairs that lead to common areas. Broken or cracked handrails must be repaired.

LOCKS

- Everything dwelling unit entrance door other than a sliding glass patio shall be equipped with a deadbolt lock with at least a one" saw resistant bolt projection, Or with a rim-mounted deadbolt lock or vertical drop deadbolt lock. The lock must be opened by turn piece, knob, or handle, which does not require a key special knowledge to operate on the inside.

FLOORS

- Replace cracked, torn, broken, or missing tiles/ linoleum.
- Replace torn, damaged, or severely worn carpeting.
- Clean dirty and stained carpet. (Inspectors will look for signs of water damage as a clue to other problems involving leaky faucets, radiators or appliances.)

WALLS

- Repair any holes and any peeling, cracking or chipping and bulging Pain/plaster.
- Repair/install baseboard/ quarter- round strip or other baseboard molding when there is gas Between the wall and floor.

SINKS

- Repair leaking faucets, clogged drains and chips in porcelain sinks.

STOVES

- Oven door and burners must operate properly.
- Gas stoves must be connected with a flex hose. A hand- controlled gas cut off valve must Be provided by the owner.
- A microwave oven can substitute for an inoperative stove, but arrangements must be made to fix stove or purchase a new stove.

CEILING

- Repair any cracked plaster or holes
- Repair water damage. investigate source of leakage and correct.

GARBAGE/DEBRIS

- Unit and yard area must be free of trash and garbage.

WATER HEATER (gas, oil electric)

- Remove all chemicals, cloth and papers stacked within 7 feet of the water heater.

- Water heaters must have a safety features: a pressure relief valve, extension pipe to within 6" of the the floor, and a hand- controlled gas cut- valve.
- Ventilation must be provided.

REFRIGERATOR

- A refrigerator must be clean and in operating condition.
- Repair/replace missing or broken shelves and kick plated.

KITCHEN

- Must have storage place for food.
- Must have a place to prepare food.

BATHROOM

- A flushing toilet that works.
- A tub or shower with hot/cold running water.

OTHER ROOMS (Rooms used For living i.e., bedrooms, dens, halls, finished basements, enclosed heated porches)

- Must have natural illumination (a window)

- Must have electric light Fixtures or outlets.

EXTERIOR OF HOME/ APARTMENT

- A roof that is in good condition and that does leak.
- Chimney (no leaning or defects such as cracks or missing bricks)

SAFETY/HEALTH

- If you reside in an apartment/ high-rise, or condo the building must: Have at least one working smoke detector on each level of the unit including the basement. have a smoke detector with an alarm system designed for hearing impaired children in care.

Exits

- Have an alternate means of exit in case of fire (such as fire stairs or Exit through window, with use of a ladder if windows are above the second floor).

ELEVATORS

- Must be safe and work properly.

ENTRANCE

- Have an entrance from the outside on from a public hall, so that it is necessary to go through anyone else's private apartment to get to the unit.

LIGHTS

- Have lights that work in all common hallway and Interior stairs.

STAIRS HALLWAYS

- Have railing on interior stairs.
- Hallway must be safe and in good condition.

POLLUTION

- No serious a pollution problems, such as exhaust fumes or sewer gas.

RODENTS and VERMIN

- No signs of rats/ mice in large number or vermin (live

WHAT'S YOUR FIRE SAFETY IQ?

Fire kills more than 5,000 people in the United States each year. Most of them in their own homes. Sadly, most of those fire death is preventable.

1. What is the leading cause of home fire?
   A. Smoking materials
   B. Arson
   C. Heating equipment
   D. Electrical equipment

2. What is the number one cause of home fire fatalities?
   A. Lightening
   B. Smoking materials
   C. Cooking materials
   D. Electrical equipment

3. Where do most of the fire death occur?
   A. Schools
   B. Homes
   C. Work
   D. Vehicles

4. How often should you replace a smoke detector?
   A. After six months
   B. After one year
   C. After the battery stops working
   D. After ten years

5. When do the largest number of home fire and associated fatalities occur?
   A. Spring
   B. Summer
   C. Fall
   D. Winter

6. Which of the following time segments accounts for the largest number of fire deaths?
   A. Midnight to 4AM
   B. 4AM to 10 AM
   C. 10AM to 6PM
   D. 6PM to midnight

7. In what room do the largest number of home fire start?
   A. Closet
   B. Utility room
   C. Kitchen
   D. Attic

8. Most fire deaths result from burns.
   A. True
   B. False

9. If fire were to occur while you were sleeping, the smoke would awake you.
   A. True
   B. False

10. If your clothes catch on fire, you should:
    A. Run to the bathtub or shower
    B. Sit still, yell for help
    C. Stop drop and roll
    D. Put baking soda on it

11. If a small grease fire starts when you are cooking, you should NOT:
    A. Slide a metal lid over the pan
    B. Extinguish it with water
    C. Slide a glass lid over the pan
    D. Turn off the heat

12. If you receive a mild grease burn while cooking, you should:
    A. Place the burn area in running hot water for about 15 -20 minutes
    B. Smear honey butter on it for about 15 -20 minutes
    C. Pour olive oil on it
    D. Place the burn area under running cold water for 15 -20minutes,

FIND A WORD FOR FIRE
SAFETY BE FIRE SMART

Complete each sentence below by choosing a word from the list. Then find each word and circle it in the puzzle.

Nine ladders match smoke stay calm cool water firefighter paramedic firecracker C.P.R firesafe smoke detector E.D.I.T.H ambulance fire engine milk and water cord stop drop roll pressure get out

1. If your clothes catch on fire, you should _____, _____, and _____
2. The number one rule in an emergency is to _____ _____
3. This will wake you up if there is a fire in your house: _____ _____
4. Put burns under _____ _____
5. If you want to stop bleeding, you should apply direct _____
6. Exit drill in the home is shortened and called ___ __ __ __ __ __
7. The first thing to doing a fire is a _____
8. A person who puts out a fire is a _____
9. If someone is very sick, you should call a _____
10. The emergency phone number is _____ one-one
11. A person who works on an ambulance is a _____
12. This kind of truck comes to your house to put out fires: _____
13. If the fire is on the second floor, you can escape with escape _____
14. This is abused on the Fourth of July and can cause serious injuries _____
15. Give these to dilute poisons _____ _____ _____
16. Playing with these, you can get burned: _____
17. If someone stops breathing and their heart stops, begin ____ ____ _____
18. Check for worn electrical _____
19. Adults should _____ bed
20. Everyone should make their home _____

45

GIVE THEM LOVE...... GIVE THEM LIFE...... GIVE A SMOKE DETECTOR

Yes __ No __ Have you removed all combustible rubbish, leaves, and debris from your yard?

Yes __ No __ Have you removed all waste, debris, and litter from your garage?

Yes __ No __ If your store paint, varnish etc. in garage, are the containers always tightly closed?

Yes __ No __ Is there an approved safety can for storing gasoline for lawn mowers, snow blowers, snowmobiles, etc.?

Yes __ No __ Do you keep your basement, storerooms, and attic free from rubbish, oil, rags, old paper, mattresses, broken furniture? Are there enough metal cans with metal covers for rubbish and combustible debris?

Yes __ No __ Are stoves, broilers and other cooking equipment kept clean and free of grease?

Yes __ No __ Are curtains near stoves arranged to prevent their blowing over the burners or flames?

Yes __ No __ Are members of your family forbidden to start fires in stoves, fireplaces, etc. with kerosene or other flammable liquids?

Yes __ No __ Are all your electrical appliances including irons, mixes, heaters, lamps, fans, radios, television sets, and other devise listed by Underwriters Laboratories, Inc.?

Yes __ No __ Do all rooms have adequate number of outlets to take care of electrical appliances?

Yes __ No __ Have you done away with all multiple attachment plugs?

Yes __ No __ Are all flexible electrical extension and lamp cords in your home in the open; none placed under rugs, over hooks, through partitions or door openings?

Yes __ No __ Do you keep matches in metal containers away from heat and away from children?

Yes __ No __ Do you extinguish all matches, cigarettes cigar butts carefully before disposing of them?

Yes __ No __ Do you know that the telephone number of the Chicago Fire Department is 911?

Yes __ No __ Do you have a plan of escape for home in case of a fire?

Yes __ No __ Do you hold fire drills in your home?

Yes __ No __ When you empty sitters, do you instruct them carefully on what to do in case of a fire?

Yes __ No __ Did your entire family take part in this checklist?

Yes __ No __ Do you have at least one (1) self-contained UL approved lifesaving smoke detector in your home?

Yes __ No __ Are all the members of the household instructed not to smoke in bed?

HOME FIRE SAFETY TIPS PLEASE USE THEM WISELY

PREVENTING BURNS AT HOME

Preventing Burns

Most burn injuries happen in the home, and most burns can be prevented.

In the Kitchen

- Make sure the handles of pots and pans do not stick out over the edges of the store where they could be bumped into.
- Do not leave stirring utensils in pots while cooking.
- Turn off all burners and oven when they are not in use.
- Have adequate dry potholders or oven mittens hung near your stove. Using a wet potholder can result in a severe steam burn.
- Do not toss wet foods into deep -fat fryers or frying pans containing hot grease. The violet reaction between the fat and water will splatter hot oil.
- Remove lids and covers from pots of cooking liquids carefully to prevent steam burns. Remember, steam is hotter than boiling water.
- Use only proper containers in a microwave oven. Let microwave- cooked food or liquids cool carefully before removing covers.

- Do not allow children or pets to play in the area where you are cooking.
- Wear tight fitting sleeves when cooking.

Hot Water

- Adjust your water heater's thermostat to below 120 degrees F to prevent scalds at the kitchen sink or in bathtubs and showers.
- Always turn on cold-water faucet first, then add hot water

Child Safety

- Keep Matches and smoking materials out of the reach of children.
- Do not allow children to play around fireplaces fires or around working space heaters.
- Cover unused wall outlets with safety caps and replace any damaged, frayed, or brittle electrical cords.
- Do not leave hot irons unattended. Do not leave hot barbeque grills unattended and supervise children cookout activities such as toasting marshmallows or hot dogs.
- Teach your children that steam radiators, stove burners, irons, and other familiar household items are sometimes hot

First Aid for Burns

Cool it: For first- and second-degree burns, cool the burned area- preferably with running cool water for 10 minutes to carry the heat away from the victim's skin and reduce pain. Third -degree burns require emergency medical treatment. Cool them only with wet sterile dressing until help arrives.

Do not use Grease: Putting butter or grease on a burn holds in heat, which makes the injury worse. Do not do it.

Wrapping Up: After cooling the burned area, wrap it loosely in sterile gauze or clean cloth.

Treat for Shock: To minimize the risk of shock, keep the victim's body temperature normal. Lay the victim on his or her back and cover unburned areas with a clean, dry blanket. Remove rings or tight clothing from around the burned area before swelling sets in and, if possible, elevate the injury areas.

TYPES OF BURNS

Every family member should know how to recognize the types of burns and knowhow to treat them. These are six major types of burns.

Flame Burns: Flame burns are caused by clothing catching on fire from a stove burner, match, candle, or other open flames.

Scalds: Caused by hot liquids or steam

Contact Burns: The results of touching hot objects.

Electrical Burns: The result of contact with uninsulated live wires or unprotected electrical outlets.

Chemical Burns: Often work related. The results of contact with a corrosive chemical, such as a battery acid.

Ultraviolet Burns: Caused by over exposure to the sun or sun lamps.

IF YOUR CLOTHING CATCHES ON FIRE, REMEMBER: STOP, DROP AND ROLL!!!

The severity of burns caused by burning clothing can be reduced by the following these three simple steps.

Stop: Do not run. Running feeds oxygen to the fire and makes it worse.

Drop: Instead, drop immediately to the floor.

Roll: Cover your face with your hands and roll over and over to smother the fire

Normally, many people, especially children panic when their clothing catches fire. Be prepared to tackle them if necessary and to help them follow these steps.

A matter of degrees

Burns are classified by the amount of damage done to the skin and other body issues.

First- degree burns: (symptom: reddened skin) are minor and heal quickly)

Second- degree burns: (symptom: blistered skin) are minor and heal quickly)

Third-degree burns: (symptom: charred tissue, often surrounded by blistered areas) are severe and require immediate professional treatment.

ELECTRICITY

Home Tips	Trouble Signs
− Do not overload sockets or extensions cords Overload outlets can overheat & cause fires.	**− Outlets that do not work**
− Repair or replace appliances immediately If they short out or spark. Hire a professional.	**− Lights, switches, wall outlets, or electrical cords that feel hot, or tingle when touched**
− Replace all frayed or cracked electrical cords They can get hot and spark causing a fire.	**− Fuses or circuit breakers that trip trip**
− Never remove a pug by pulling or yanking on it Always remove by holding the plug	**− Lights that flicker dim**
− Do not place cords under carpet's Tape them to the wall where they will not be walked on	**− Unusual smells that cannot be accounted for in the house**
− Have a professional check Your own wiring projects	

Install smoke detectors on each level of your home

Check them once a month and change their batteries at least once a year

THERE ARE CLEAR SIGNALS TO HAVE THE WIRING IN YOUR HOUSE CHECKED BY A PROFESSIONAL.

WHEN A FIRE OCCURS

ALWAYS:

- ✓ First notify 9-1-1; sound any alarms
- ✓ Evacuate immediately, no fire to small
- ✓ Rescue anyone in danger
- ✓ Be aware of directions prior to use
- ✓ Use only on small and confined fires
- ✓ Keep a clear escape route; do not let the fire get between you and the exit
- ✓ Stay low, below the smoke
- ✓ Leave the fire if it grows out of control
- ✓ Close doors to confine the fire as you leave
- ✓ Wait until the fire department inspects the area before re-entering the building

NEVER

- ✓ Never start reading directions
- ✓ Never use on large or growing fires
- ✓ Never fight fires without an escape route
- ✓ Never fight fires in smoke filled rooms
- ✓ Never fight fires if you are in doubt
- ✓ Never assume the fire is out until the area has been inspected by the fire department

FIRE SAFETY TIPS

Smoking Materials:

Cigarette, matches, lighters, and candles are the leading cause of fires!

- ✓ Smoke was permitted
- ✓ Use deep, large non-tip ashtrays
- ✓ Make sure the ashtray contents are cold before emptying
- ✓ Do not waste baskets for ashtrays

Electrical Wiring:

- ✓ Replace electrical cords that have cracked insulation or a broken connector
- ✓ Do not exceed the amperage load specified for extension cords
- ✓ Get rid of multi-plug attachments
- ✓ Do not run extension cords across doorways or under carpets, where they can be stepped on and damaged

Electrical Appliances:

- ✓ Leave plenty of air space around all machines (computers, copy machines, etc.)
- ✓ Keep heat appliances away from materials that could ignite (coffee makers, toasters)
- ✓ Designate someone to unplug all appliances at the end of the day, or when not in use.

HOME FIRE SAFETY CHECKLIST

- Use large ashtrays and make sure all smoking materials are out before getting rid of ashes
- Install and properly maintain smoke detectors throughout your house. Assess your smoke detectors periodically.
- Keep an ABC dry chemical fire extinguisher handy and know how to use it. Follow manufacturer's inspection recommendations and recharge after each use.
- Develop a family fire escape plan and practice it every few months.
- Clean the rubbish, old papers, oily rags, magazines, and furniture out of your basement, attic, and garage.
- Keep gasoline and other flammable liquids stored in approved containers, in the garage outside your home.
- Use the proper fuses in your fuse box
- Check the insulations on all electrical cords. Also check plugs and receptacles.
- Keep matches and lighters safely stored and away from small children.
- Keep your chimney clean and use a fire screen with your fireplace.
- Space heaters must be well shielded and well ventilated. Keep them out of traffic paths and away from furniture and curtains.

PUT THE LID ON KITCHEN FIRES

How to prevent cooking fire- and how to fight them if they get started!

➢ Pay attention to your cooking. Do not overheat grease. Watch grease overflows that can start fires. If you must leave the stove to answer the phone or doorbell, turn down the heat. If you will be gone more than a few minutes, turn it off.

➢ If your children help you cook, make them aware of cooking hazards. Turn skillet and hot handles towards the center of the stove to prevent accidental overturning.

➢ Do not leave towels or napkins on or near the stove. Do not wear frilly garments-especially those with loose, floppy sleeves-while cooking.

➢ Keep a class ABC fire extinguisher in or near the kitchen. (An ABC rating indicates the fire extinguisher can be used on fires involving grease, paper towels, electrical appliances and other materials commonly found in the kitchen)

➢ If the grease fire is small, you may be able to stop it with a handful of baking soda (bicarbonate of soda). But do not use baking powder, which contains flour or starch, it could spread the fire. And never use water on grease fire, it also increases the chance of this type of fire getting out of hand.

➢ Always have a pot lid handy to smother a small grease fire

➢ Do not try to move or carry a pan that is on fire from grease. Even though moving the pan is common reaction when a grease fire is discovered, it often results in burns to the carrier and additional fire damage.

➢ If a fire is a big one, do not try to fight it-call the fire department.

Cooking....

Especially when it involves grease, is one of the leading causes of fire in the home. And while kitchen fires seldom kill people, they injure thousands and cause property damage in hundreds of millions of dollars each year.

SUMMER SAFETY
HEAT STRESS

Heat related deaths and illnesses are preventable. Annually many people succumb to the heat.

Signs and Symptoms of heat stress

- Elevated body temperature
- Muscle cramps or clammy skin
- Headaches and nausea
- Dizziness and/or fainting

Reduce your risk of heat stress

- Increase your fluid intake
- Avoid caffeinated drinks
- Wear loose and breathable clothing
- Take short breaks in shaded areas

SUMMER BBQ SAFETY TIPS

Make sure your grill is clean thoroughly.

Set the Grill (Charcoal or Gas) up safely before starting.

Always set up the grill 10 feet away from any structures: Buildings, Garages, or Shed.

Place on a flat surface, away from buildings, trees, or shrubs.

Preheat the equipment, making sure all food is thawed and hands are clean.

Keep chilled food cold, Avoid cross-contamination, ensure food is cooked all the way through, and serve food immediately, especially for those who have the stomach music playing.

NEVER LEAVE THE GRILL UNATTENDED OR ALLOW CHILDREN TO MANAGE!

INSURANCE

What is insurance and how does it work?

Throughout the years people have used insurance to protect themselves from risk due to fire and other catastrophes. The types of insurance are life, auto, health, and home insurance. Some of the risk that people experience is:

- Personal Risk- Catastrophes that effect individuals (i.e., illness disability)
- Property Risk-Property that is damaged or destroyed (i.e., natural disaster fire)
- Liability Risk- Events that effect the person or property of others which could result in injury (i.e., a visitor gets hurt in your home)

The way insurance works, a large group of individuals pay a yearly premium that is entered into a common fund. When disaster strikes one member of the group pays for the loss, this process shifts to a different member each disaster.

HOME INSURANCE

A home is one of the largest expenditures in most people budget. Furniture and other personal belongings represent a sizeable investment. Therefore, people protect their home and its contents form damage or loss.

There are various kinds of homeowner insurance, whether you live in a house, apartment, mobile home, or condominium there is a policy designed for your dwelling. Good homeowners' insurance makes sense, but it is wiser to know how your policy works.

Landlord- Tenant Relationship

The relationship between you and your property owner is a legal one. In this relationship you both have certain rights and responsibilities as it relates to safety.

The rights and responsibilities of the property owner are as follows

- Has the right to set reasonable rules and regulations for the management of the rental units (unless you have agreed to repair or maintain the property)

- Has the right to enter the apartment for inspection or maintenance. Note: Tenants should approach the property owner and request that prior notification is provided (except for emergency's) for future inspections of the apartment.
- Has the right to keep the place in reasonable repair
- Has the right to keep premises clean and in safe condition
- Has the right to receive rent for housing
- Is responsible for providing a safe and livable apartment
- Is responsible for fulfilling his obligations
- Is responsible for having a way to be contacted in case of and emergency (i.e., a telephone number, through an assistant

The rights and responsibilities of the tenant is as follows:

- Is responsible for paying rent (depending on the arrangement (i.e., section eight) on time
- Has the right to be provided with peace, quiet and privacy
- Has the right to not be cheated or over charged
- Is responsible for following the rules of the lease
- Is responsible for not abusing the property
- Is responsible for keeping their area clean and safe
- Is responsible for not abusing the rights of the tenants
- Is responsible for having in writing all communication with the property owner and always make copies of documents signed or received

As a tenant you should try to maintain a professional relationship with the property owner. If you see a safety problem, report it as it happens, hopefully the property owner will make provisions to rectify the situation right away (if possible). If you leave your apartment to go on a vacation or just for a few days, it might be a good thing to inform your property owner and leave him/her information on how you can be contacted in case of an emergency.

Unfortunately, not all property owner/tenant relationships work, but for the relationship to remain professional both parties must fulfill, understand, and carry out their obligations.

Safety is your right, exercise it!

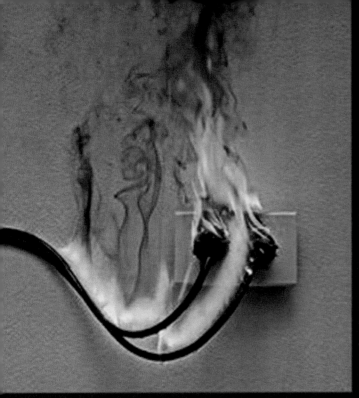

EXTENSION CORD

Smoke or burn marks on an outlet are often caused by heat resulting from an overloaded circuit, however, there are several other things that should be checked as well. Smoke or burn marks should be noted as a red flag that there may be a fire safety concern.

➢ Arcing

➢ When an electrical current leaps across a gap, this intermittent connection causes a spark or arcing which translates to heat.

The source of the heat causing the outlet face to be burned or show smoke can be due to several causes, including the outlet having the hot (energized) wire touching the neutral or ground, or it just being close enough for arcing to occur. Loose wire connections that are not secured down tightly can also be at fault.

Extension cords are only supposed to be temporary measures. You are inviting trouble if you try to use them long-term to load 4 or 5 appliances onto one outlet.

You should also avoid running extension cords under rugs, particularly in high-traffic areas. As people walk over the rugs, they often fray the extension cord's insulation, which is just asking for a fire. But you might never know it's happening, because it's "out of sight, out of mind."

ELECTRIC BLANKET FIRE CAUSES

The electric blanket can be a useful, convenient tool for consumers. However, the use of an electric blanket can also increase the risk of a house or property fire, and of personal injury. Due to the nature of the product, electrical fires, smoldering, and full flame combustion are possible, and the chances of these incidents occurring increases greatly when the product becomes worn or damaged.

Experts estimate that an average of 5,000 house fires are caused by electric blankets every year. These fires typically occur due to any one of the following reasons:

- Manufacturing mistakes; i.e., improperly installed wiring, faulty control unit. This cause is rather uncommon, and the least often occurring cause in cases of electric blanket fires.
- Improper handling of the electric blanket unit, such as prolonged use or unsupervised use in situations concerning small children or elderly persons with physical handicaps.
- Normal wear and tear, which can cause the wire implements or other components to fray, break, or become otherwise damaged and susceptible to malfunction.

It is believed that 99% of all fires and other accidents related to electric blankets and heating pads are caused by units which are 10 or more years old. Regular safety checks and efforts to discard and replace old blankets can help to prevent fires and injuries from occurring.

Types of accidents caused by electric blankets:

- 12% Heat contact
- 12% Electrocution
- 18% Smoldering
- 58% Caught fire

ELECTRIC BLANKET INJURIES

Injuries and their percentages caused by electric blankets:

- 2% Electric burns
- 37% Burns, minor
- 60% Smoke inhalation

Electric blanket malfunctions can cause personal injury or damage to property. Smoldering, sparking, scorch marks, and missing components can all be signs of an unsafe electric blanket. If you notice or suspect any of these characteristics to be occurring in your own electric blanket, discontinue use and discard the blanket immediately.

When a fire destroys property and causes injury because of a manufacturer defect or a failure of the manufacturer to provide clear warnings about the product's use, the victim is entitled to restitution. A personal injury lawyer can handle your case and negotiate to win you a settlement that is fair and sufficient to cover your costs of recovery.

ELECTRIC BLANKET FATALITIES

Fatalities caused by electric blanket related accidents:

- 1% Various/unreported ages
- 1% Children under 5 years
- 9% Individuals between the ages of 18 and 65
- 89% Individuals over the age of 66

Fatalities due to electric blanket fires are rare, but do occur. Along with normal precautions and safety assessments, consumers should be certain to never leave a child, an elderly individual, or any individual having or experiencing sensory limitations with an electric blanket for extended, unsupervised periods of time, or to leave an electric blanket plugged in and powered on overnight.

If you or someone you know has experienced property damage, injury due to electric blanket fire, smoldering, or electrocution, or if someone you know has died due to electric blanket malfunction, please do not hesitate to contact the McLaughlin Law Firm, P.C., at (720) 420-9800.

WHAT IS A SPACE HEATER?

Space heaters effectively provide heat to warm up a room using different technologies. They can't heat up connected rooms, unlike central heating systems – they're good for closed spaces. In simple terms, they are the solution to make a room comfortable and keep winter chills at bay.

Space Heater Safety Tips

- Keep flammable stuff like drapes and clothes away from the device.
- Don't keep anything too near the heater – maintain distances of four feet.
- Don't fall asleep while the heater runs.
- Try not to leave the room for too long when the heater is running.
- Avoid using extension plugs – use the wall outlet.
- Keep it at a place where it won't topple over.
- Any source of moisture should be kept away from space heaters.
- Check to make sure your smoke alarms work properly.
- Keep it away from the reach of pets and children.
- Make sure it's in good condition – clean filters and no damaged cords.

Types of Space Heaters

Ceramic - Ceramic space heaters can be radiant or convective. Inside a ceramic heater, several ceramic plates are present. It uses electricity to heat these plates to the extent that they start radiating heat. After this, it spreads the heat by radiation or forces it out with a fan.

Oil-Filled -These heaters work in almost the same way as a ceramic heater – just replace the ceramic plates with oil.

Infrared - infrared space heaters emit electromagnetic waves to heat objects and living beings nearby. It's a similar process to how the sun heats up our planet. Hence, infrared heaters can't heat up the entire room. They're the best type of heaters for spot heating – if you're in the line of sight, you'll feel instant warmth.

Propane - Unlike the rest of the types we've discussed, propane space heaters don't use electricity to run. Instead, they utilize liquid propane to operate. This means that they combust fuel and release fumes for which they need venting. However, some models are vent-free for your convenience.

Lead is a highly toxic metal that produces a range of adverse health effects particularly in young children.

Where Is It Found?

There are many ways in which humans are exposed to lead: through deteriorating paint and dust air, drinking water, food, and contaminated soil. Airborne lead enters the body when you breathe or swallow lead particles or dust once it has settled. Lead can leach into drinking water from certain types of plumbing materials (lead pipers, copper pipes with lead solder, and brass faucets) and can also be found on walls, woodwork, and the outside of your home in the form of lead-based paint. Lead can be deposited on floors, windowsills, eating and playing surfaces, or in the dirt outside the home.

About two-thirds of the homes built before 1940, and one-half of the homes built from 1940 to 1960 contain lead-based paint. Some homes built after 1960 but before 1978 may also contain lead paint. Most paint made after 1978 contains no intentionally added lead, since it was banned from use on the interior and exterior of homes.

Even though leaded gasoline is seldom used today, high levels of lead found in soil can be attributed to past emissions.

Children can swallow harmful amounts of lead if they play in the dirt or in dusty areas (even indoors) and then put their fingers, clothes, or toys in their mouths, or if they eat without first washing their hands.

What Are the Health Effect?

Exposure to excessive levels of lead can cause brain damage; affect a child's growth; damage kidneys; impair hearing; cause vomiting, headaches, and appetite loss; and cause learning and behavioral problems. In adults, lead can increase blood pressure and can cause digestive problems, kidney damage, nerve disorders, sleep problems, muscle and joint pain, and mood

The most accurate way to determine if your home has lead-based paint is to hire a lead inspector to test the paint. Lead inspectors use XRF (x-ray) instruments to determine content of lead in paint immediately. Another way is to hire a risk assessor who will take samples from several locations in your home and have them analyzed at a lab for lead content. If an individual is concerned about a specific area in a home and wants to take a simple paint chips, dust, or soil sample themselves, they can mail the sample directly to a certified laboratory and have it analyzed. Taking a sample without an assessor is easy and may be less expensive, but it only tests the area from which the paint, soil, or dust sample was taken. A house may contain several layers of pain from different periods so one or two samples may not be representative of the entire residence.

The Environmental Protection Agency has not approved and does not recommend do-it-yourself lead test kits. These kits, which do not require lab analyzation, are not very accurate in determining the existence of lead paint. For more information, or to locate lead-based paint inspectors, risk assessors and certified laboratories call (800) 424-5323.

How Can I Reduce Lead Exposure?

- If your home has lead paint, do not try to remove the lead from your home yourself. Improper removal often makes the situation worse. Hire a qualified contractor to do the work. In some states, landlords may be required by law to remove lead-based paint from homes where children have been poisoned. Check with local health officials. To locate trained lead service providers, including lead-based paint inspectors, risk assessors and abatement (lead removal) contractors in your area, call (312)747 - LEAD.
- Since lead can come from the solder or plumbing fixtures in your home, water from each faucet should be tested. Call the EPA Safe Drinking Water Hotline (800) 426-4791 for information on laboratories certified to test for lead.
- Mop floors and wipe window ledges and other areas with soapy water. If available, tri-sodium phosphate or lead-specific cleaning products can be used.
- Keep the areas where children like to play as clean and dust free as possible.
- Keep children away from areas where paint is chipped or peeling. Stop children from chewing on windowsills oar other painted surfaces.
- Make sure everyone washed their hands before meals, naptime, and bedtime.
- If your child's bottle or pacifier falls on the floor, wash it before giving it back to your child.
- Wash toys, stuffed animals, and bedding regularly.
- Send children and pets to a relative's or neighbor's house if you plan to renovate your house. Infants, children, and pregnant women should not be in the home while renovations are under way. Exposure to lead dust is hazardous.
- If you are pregnant, take as much care to avoid exposing yourself to lead as you would for your child. Lead can pass through your body to your unborn baby and cause health problems.
- Do not let your children eat sand, dirt, or paint chips. Encourage your children to play in grassy areas of the yard or playground. Plant grass in areas where children play if possible. Make sure children remove and wipe their shoes and wash their hands whenever they come inside after playing outdoors.

- Try to make sure your children eat a balanced diet with plenty of foods that contain iron and calcium. A child who gets enough of these minerals will absorb less lead. Foods rich in iron include eggs; lean red meat; and beans, peas, and other legumes. Dairy products such as milk, cheese, and yogurt are also recommended for their high calcium content.
- Do not store food or drink in containers made from crystal, because some crystal contains lead.

What Is the Residential Lead-Based Paint Hazard Reduction Act?

The Lead-Based Paint Hazard Reduction Act of 1992, known as Title X, requires that most home buyers and renters will receive known information on lead-based paint hazards during sales and rentals of housing built before 1978. Sellers and landlords are required to provide a lead-based paint disclosure form and a federal pamphlet, title *Protect Your Family from Lead in Your Home*, to the buyer or renter before the sale or lease of certain property. Landlords are also required to disclose information regarding lead-based paint to pre-existing tenants if the property was built prior to 1978 Congress passed Title X to protect families from exposure to lead by requiring disclosure of lead-based paint hazards in residential property. Title X became effective for all residential property built before 1978 on December 6, 1996.

For More Information

Contact the National Lead Information Center at (800) 424-5323 or http://www.epa.gov/opptintr/lead/nlic.htm

ENVIRONMENTAL TOBACCO SMOKE

What Is It?

Environmental tobacco smoke (ETS) is a mixture of particulars that are emitted from the burning end of a cigarette, pipe, or cigar, and smoke exhaled by the smoker. Smoke can contain any of more than 4,000compounds, including carbon monoxide and formaldehyde. More than 40 of the compounds are known to cause cancer in humans or animals, and many of them are strong irritants. ETS is often referred to as "secondhand smoke" and exposure to ETS is often called "passive smoking."

What Are the Health Effects?

Secondhand smoke has been classified as a Group A carcinogen by the U.S Environmental Protection Agency (EPA), a rating used only for substances proven to cause cancer in humans. A study conducted in 1992 by the EPA concluded that each year approximately 3,000 lung cancer deaths in nonsmoking adults are attributable to ETS. Exposure to secondhand smoke also causes eye, nose, and throat irritation. It may affect the cardiovascular system and some studies have linked exposure to secondhand smoke with the onset of chest pain. ETS is an even greater health threat to people who already have heart and lung illnesses.

Infants and young children whose parents smoke in their presence are increased risk of lower respiratory tract infections (pneumonia and bronchitis) and are more likely to have symptoms of respiratory irritation like coughing, wheezing, and excess phlegm. In children under 18 months of age, passive smoking causes between 150,000 and 300,000 lower respiratory tract infections, resulting in 7,500 to 15,000 hospitalizations each year, according to EPA estimates. These children may also have a buildup of fluid in the middle ear, which can lead to ear infection. Slightly reduced lung function may occur in older children who have been exposed to secondhand smoke.

Children with asthma are especially at risk from ETS. The EPA estimates that exposure to ETS increases the number of asthma episodes and the severity of symptoms in 200,000 to 1 million children annually. Secondhand smoke may also cause thousands of non-asthmatic children to develop the disease each year.

What Can Be Done to Reduce Exposure to ETS?

- Do not allow smoking in the home, especially around children. Do not allow babysitters and others who work in the home to smoke in the home or near your children. If someone does smoke at home, increase ventilation in the area where smoking takes place.

69

- Make sure that any outside group that assists in the care for children, such as schools and daycare facilities, has a smoking policy in force that protects children from exposure to ETS.
- If your workplace does not have a smoking policy that protects nonsmokers from exposure to ETS, try to get it implement one. See if it will either ban smoking indoors or designate a separately ventilated smoking room that nonsmokers do not have to enter as part of their work responsibilities.

Homeowners and renters may harbor deadly household items without realizing the potential fire hazard(s).

Items that have safety problems:

- Cigarette lighters without child resistant mechanism
- Space heaters
- Furnaces
- Frayed or cut extension cords
- Black & decker optima toaster sold from 1994-1996
- K-mart children's decorative lamps sold from 1993-200
- Halogen torchiere floor lamps produced in 1997 with no wire guard
- GE and Hotpoint dishwashers made 1983-1989
- Old electric hair dryers without built in shock protection devices

Please do not attempt to sale or donate any of these products. Any of the products should be taken to a fire station near you.

REFERENCES

Allstate Insurance Company, 2000, No Author (Video tape), Be Cool About Fire Safety.

Chicago Housing Authority, Guidebook, 4-1-98, Home Safety Items, Office of Fair Housing and Equal Opportunity.

Jacobs, Jodie, 1999, A Healthy Child Needs a Healthy Parent, Chicago Tribune, Records Display.

Public Education Office for Fire Prevention, 2001, Chicago Fire Department.

Public Education Office for Fire Prevention, 1998, Winston Salem, N.C.

Rodriguez, Alex, 2000, House Fire Claims Second Child's Life, Chicago Tribune. Record Display.

South Wheaton Educational Publishing Company, 1997, Working Learning and Living Standards for Firefighter Professional Qualifications, Essentials of Firefighting, NFPA.

VanBuren, Abagale, 2000, Playing with Fire is a Deadly Game, Chicago Tribune, Records Display.

APPENDICES

THE BEFRIEND
A FIREMAN
PROGRAM, BY THE
OFFICE OF FIRE
PREVENTION, INC.

The Befriend a Fireman Safety Advocate program wants you to become a Safety Advocate (SA) today. Become a Safety Advocate by making a friend request to my Facebook Safe book page and expressing your interest to Befriend Fireman and become a Safety Advocate (SA).

The next step will be a visit to your local Fire Department (as seen here by the CEO of OFP) befriend a Firefighter (say hello), shake their hand and share with them your home safety plan. The purpose of this meeting is to start the conversation about home safety and to build a fire safety relationship. Please be ready to listen and have a positive attitude.

If the Firefighter has no objections about taking a picture with you and of the handshake and if you have no objections about taking a picture with the Firefighter and of the handshake and then submitting your picture to my Safe book page, I will have no objections about posting it.

The objective of this effort is to save lives, property, money and time by doing positive things in advance that make your home and this world a safer place to live, work and play.

A Safety Advocate (SA) like you deserves to be acknowledged and recognized today. Now that you have done your part that is it, you are now a Safety Advocate (SA).

OFP and CFD proving that safety relationships are possible, **positive** and necessary

Congratulations

on

"Befriending a Fireman"

Safety Advocate

You are officially in, you have completed your first task. Now is where the serious fun begins. Your ongoing responsibilities requires that you think and know safety first in all that you do. Safety First is not just a slogan, it's a safety attitude.

Derrick A. Whitehead

Chief Executive Officer

ATTENTION

THIS HOME IS SAFETY CERTIFIED AND PROTECTED BY THE TOOLS OF PREVENTION THAT ARE INSIDE.

THE OFFICE OF FIRE PREVENTION, INC. IS WORKING TOGETHER WITH THE COMMUNITY IN AN EFFORT TO SAVE LIVES.

WWW.OFFICEOFFIREPREVENTION.COM

DERRICK A. WHITEHEAD
OFFICE OF FIRE PREVENTION

SIGNATURE

DATE

DISCLAIMER: Safety assurance after the certification date is the responsibility of the above signed.

78

The community Fire, Hazard and Prevention project is a community care program committed to recruiting advocates for safety, promoting awareness, preventing fires and hazards. The members of this program recognize that there is a need for a more vigorous approach in fire prevention and awareness, therefore we must all do our part to protect our families and community and prepare for prevention.

My name is Derrick Whitehead, I am the Chief Executive Officer of the Office of Fire Prevention, Inc. Because your dwelling is safe from hazards and equipped with the tools of fire prevention, you have received this safety flyer. The recommended placement for this safety flyer is inside the dwelling in a window for public view. The flyer purpose is to express to the community your safety operations for fire, hazard and prevention. It is important that you and your family continue to exercise safety and upkeep the proper operation for the tools of prevention.

The tools of prevention include a smoke detector, surge protector, Carbon monoxide detector, fire extinguisher and a sprinkler system. In the event that you do not have all the tools to prevent, fire and hazards, you are encouraged to invest in as many as possible. By law you must have a smoke detector and a carbon monoxide detector in your dwelling.

It is very important that you and your family understand how the tools of prevention work. I encourage you to contact your local Fire Department or myself if you have questions or concerns regarding operations. Thank you for being an advocate for safety in your home and community by making an effort to promote, protect and prevent fires and hazards.

OFFICE OF

FUEL HEAT OXYGEN

FIRE PREVENTION, INC.

Derrick A. Whitehead
Chief Executive Officer
Certified Firefighter

OFFICE OF FIRE PREVENTION, INC.

MONTHLY SAFETY CALENDAR

"WORKING TOGETHER TO SAVE LIVES THROUGH DETECTION WHICH INCREASES THE CHANCE OF PROTECTION".

MONTH/YEAR _____

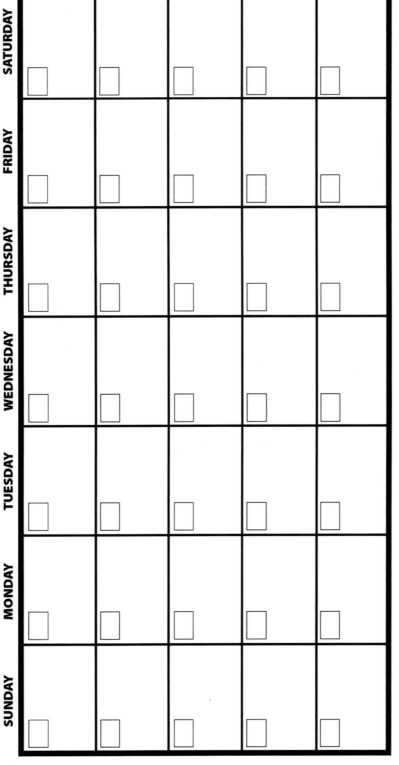

SUNDAY	MONDAY	TUESDAY	WEDNESDAY	THURSDAY	FRIDAY	SATURDAY
☐	☐	☐	☐	☐	☐	☐
☐	☐	☐	☐	☐	☐	☐
☐	☐	☐	☐	☐	☐	☐
☐	☐	☐	☐	☐	☐	☐
☐	☐	☐	☐	☐	☐	☐

SMOKE DETECTOR
SYSTEM TEST ___ / ___ / ___
BATTERY TEST ___ / ___ / ___
BATTERY CHANGE ___ / ___ / ___

FIRE EXTINGUISHER
PURCHASE DATE ___ / ___ / ___
LAST USE ___ / ___ / ___
CLASS _____

CARBON MONOXIDE DETECTOR
DISPATCH TEST ___ / ___ / ___
BATTERY TEST ___ / ___ / ___
BATTERY CHANGE ___ / ___ / ___

EVACUATION PLAN
PRACTICE DATE ___ / ___ / ___
PARTCIPANTS ___ / ___ / ___

NOTES: _____

"Test your smoke detectors once a month, following the manufacturer's instructions"

HOLIDAY SEASON SAFETY TIPS

The winter holidays have arrived and these are the days when family, friends and neighbors come together to celebrate the season.

The Office of Fire Prevention, Inc.

wants to encourage you to be safe because with the winter holiday season comes greater risk of fire, hazards and accidents. So in your efforts to prepare for the holidays, please practice prevention by putting safety first.

Space Heaters

- Keep flammable stuff like drapes and clothes away from the device.
- Don't keep anything too near the heater – maintain distances of four feet.
- Don't fall asleep while the heater runs.
- Try not to leave the room for too long when the heater is running.
- Avoid using extension plugs – use the wall outlet.
- Keep it at a place where it won't topple over.
- Any source of moisture should be kept away from space heaters.
- Check to make sure your smoke alarms work properly.
- Keep it away from the reach of pets and children.
- Make sure it's in good condition – clean filters and no damaged cords.

Celebrating

- Have your home fire escape plan in an inconspicuous location of the home and inform your guest that you have a plan and where it's located.
- When cooking on the stove in an effort to prevent a fire, stay in the kitchen.
- Have a plan for smokers to prevent fire.
- Check cords for loose bulb connection for tightening or replacement.
- Always read the manufacturer's instructions for proper use of the product before using.

Tools of Prevention/Detection

- Test you Smoke Alarms
  Change the batteries if necessary
- Test your Carbon-Monoxide Detector
  Change the battery if necessary
- If your home has a Fire Extinguisher
  Know the directions for reuse before its necessary
- If your home has a Surge Protector or Extension Cord(s), do not over-load them by plugging heavy duty appliances into them. Avoid running extension cords under rugs, particularly in high traffic area.

Decorating

- When choosing decorations, choose decorations that are flame resistant or flame retardant.
- If candles are a part of your decorations, keep lit candles away from anything that can burn.
- Keep children and pets away from lit candles
- Blow out all lit candles when you go to bed or leave the room especially if children are in the area.
- Replace any string lights with worn cords or broken cords.
- Check cords for loose bulb connection for tightening or replacement.
- Always read the manufacturer's instructions for proper use of the product before using.
- Turn off all lights and decorations when going to bed.

82

FROM THE OFFICE OF FIRE PREVENTION, INC. FAMILY TO YOURS, HAPPY HOLIDAYS TO ALL AND TO ALL A SAFE HOLIDAY SEASON

CHECK OUT MY WEBSITE FOR MORE SAFETY TIPS

http://www.officeoffireprevention.com/

SA f E BOOK

Office of Fire Prevention

@Officeoffireprevention · Fire Protection Service

My name is **Derrick Whitehead.** I am a native of Chicago, Illinois. I grew up in the Englewood area. I have an undergraduate degree from "Winston Salem State University (WSSU)" in Winston Salem, North Carolina, in Business Administration. I have an MBA/ Graduate Degree, in Human Resources from Dominican University in River Forest, Illinois.

I received my certification to fight fires in Winston Salem, North Carolina, Firefighter Level II, Hazmat Level I. It was during my experience as a Firefighter where I came up with the idea to start the Office of Fire Prevention, Inc. (OFP).

It was an early AM before my shift ended in December, when our engine company was called to a three-alarm fire. Upon our arrival, the home was engulfed in flames and an audience of community members. I was the second man on the hose and our first responsibility upon entering the structure was to initiate the search and rescue efforts.

As we exited the truck to position ourselves on the hose, I am thinking, "this is an intense burn, I hope that no one is in there", then I hear the crowd of onlookers yelling "there is a child inside, there is a child inside". After receiving directions from the captain, it was our time to go in. We entered the home through the front door and safely made our way up the steps to the second level of the home, touching all doors, looking under beds and looking in closets in search of life.

Our initial search efforts did not find any victims but a second engine company search efforts located a body on the first level. After our efforts on the second level, my team proceeded to the first floor; it was then that I saw something, which at the time I did not know would change my life forever. It was the body of a 10-year-old male. The child laid there unresponsive and burned badly. Over half of this child's body was burned and his teeth were melted from the intense heat of the flames. My immediate reaction after observing the body was extreme sadness, devastation and trauma. I had never seen anything like this before. I immediately thought about the people that I love, thinking to myself "the 1st opportunity I get after finishing up on this fire scene I was going to call them". I was going to call my daughter, and next, my mother and then, my best friend and tell them all how much I love them and to make certain that they make sure their homes are fire safe with a fire extinguisher and working detectors (smoke, surge and carbon monoxide detectors). As I continued to work the fire scene, my anxiety was getting the best of me. I could not wait to make the calls and talk with them about the importance of good housekeeping, making sure that they do not have items obstructing their means of entrance and or egress, drafting an escape plan and practicing it, checking all windows to ensure that they are not nailed or painted shut and more.

Based on what I experienced, I wanted them to know how it inspired me to want to do more to promote fire prevention and save lives, time and property from the havoc of fire and hazards. The home was the poster home for unsafe fire conditions. The home did not have any working smoke or detectors in it, it used extension cords, which were over loaded, and from the looks of all the clutter in this home it appeared as if someone in this family was a hoarder.

This experience illuminated my passion and commitment to taking fire safety and its message to another level and that is when I began to work on what today is now known as the Office of Fire Prevention, Inc. (OFP). I personally pledged to do more as a firefighter fighter and a citizen to promote prevention and safety to all communities.

Its 25 years later and today, my services includes trainings for businesses, schools, residential facilities, senior homes, Public Housing and more. I have community care programs, a Safety line of clothes, Safety Products, Safety Slogans and a stronger ambition to work together with the community and its stakeholders to promote fire safety awareness. My vision is to not only educate the public about fire prevention awareness, but to actuate them. While it is a good thing to know what to do in the event of a fire, it is a better thing to show that you know what to do before a fire starts by properly practicing prevention. Knowing what to do in a fire does not save your life, preparing for a fire in advance will.

The efforts of the Occupational Safety and Health Administration, (OSHA), National Fire Protection Association (NFPA), the Office of Fire Prevention, Inc. (OFP) and your local Fire Department are futile without your actions. In short, we need you in order to do our jobs successfully.

The Office of Fire Prevention, Inc. mission is to strengthen the nation's knowledge about fire safety, fire prevention and fire hazards. OFP is committed to protecting people from risk, injury and danger and we plan to accomplish this by making safety a priority as well as a practice for everyone, in their work, life and play, so please join our team by becoming a Safety Advocate today.

The Twelve P's of Fire Safety

Pit Poor Preparation, Promotes Pit Poor Performance, Pit Poor Prevention Provokes Panic